# Teach Me the Water Cycle and Weather

# Coloring Book

# Tamika K. Fordham

Copyright © 2012 Author Name

All rights reserved.

**Title ID: 7617363
ISBN-13: 978-1981965977**

DEDICATION
THIS BOOK IS DEDICATED TO MY SON
KENNEDY FORDHAM. THANK YOU FOR BEING
SUCH A SMART, INTELLIGENT, AND CREATIVE
SON. YOU ARE MY LITTLE SCIENTIST!

ACKNOWLEDGMENTS

TO MY MOTHER WHO ALWAYS TAUGHT ME TO GET A GOOD EDUCATION AND ALWAYS PUT GOD FIRST.

About 71% of the Earth's surface is covered with water. Oceans hold 95% of all the water on Earth.

Liquid water on Earth becomes a gas when heated, called water vapor. This process is called evaporation. The process of evaporation results from the Sun's heat energy.

Condensation happens in the air as water vapor changes back to droplets of water. Clouds form as a result of condensation. Dew also forms from condensation, but the water droplets condense directly onto a surface such as grass, a school bus, or windows. The process of condensation results from the cooling of the air temperature.

After condensation occurs (allowing for the forming of clouds), any form of water that falls from the clouds is called precipitation (rain, snow, sleet, or hail). Snow, sleet, and hail result from freezing temperatures 32 degrees and below in the air inside of clouds. Rain forms when the air temperature is above freezing temperatures.

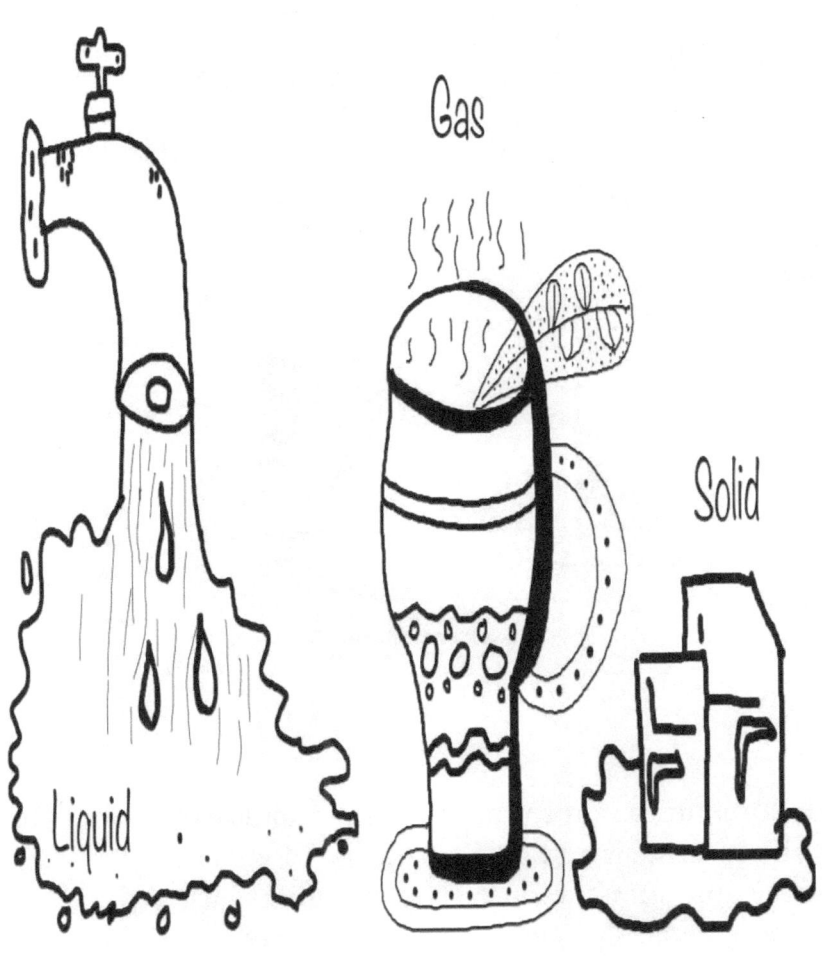

It sometimes rain for days during a stationary front. That's when warm air and cold air meets. Floods occur when a large amount of water covers land that is usually dry. Floods typically occur due to large amounts of rainfall. A steady light drizzle occurs during a warm front. That's when warm air move over cold air.

A cold front is when cold air move over warm air. It creates thunderstorms. Thunderstorms are severe storms with lightning, thunder, heavy rain and strong winds. Some thunderstorms produce hail. Some examples of the effects of thunderstorms may be that heavy rains cause flooding, lightning can cause fires, and strong winds can blow over houses and trees. During a thunderstorm, stay inside; stay way from water; and do not stand near tall objects.

A funnel-shaped cloud that comes down from a storm cloud with winds spinning at very high speeds. Some examples of the effects of tornadoes may be that high winds damage and destroy houses, cars, and trees. During a tornado, find a safe place away from windows; if you cannot find shelter lie flat in a ditch and do not stay in your car. The strong winds will move under it and lift the car into the air.

A Hurricane is a large storm that forms over warm ocean water with very strong winds that blow in a circular pattern around the center, or eye, of the storm. Some examples of the effects of hurricanes may be that high winds can blow over trees and buildings; heavy rain can cause flooding; the ocean waves and rise in sea level can cause massive flooding and damages. During a hurricane, board up your windows; and move further inland if you are near the coast.

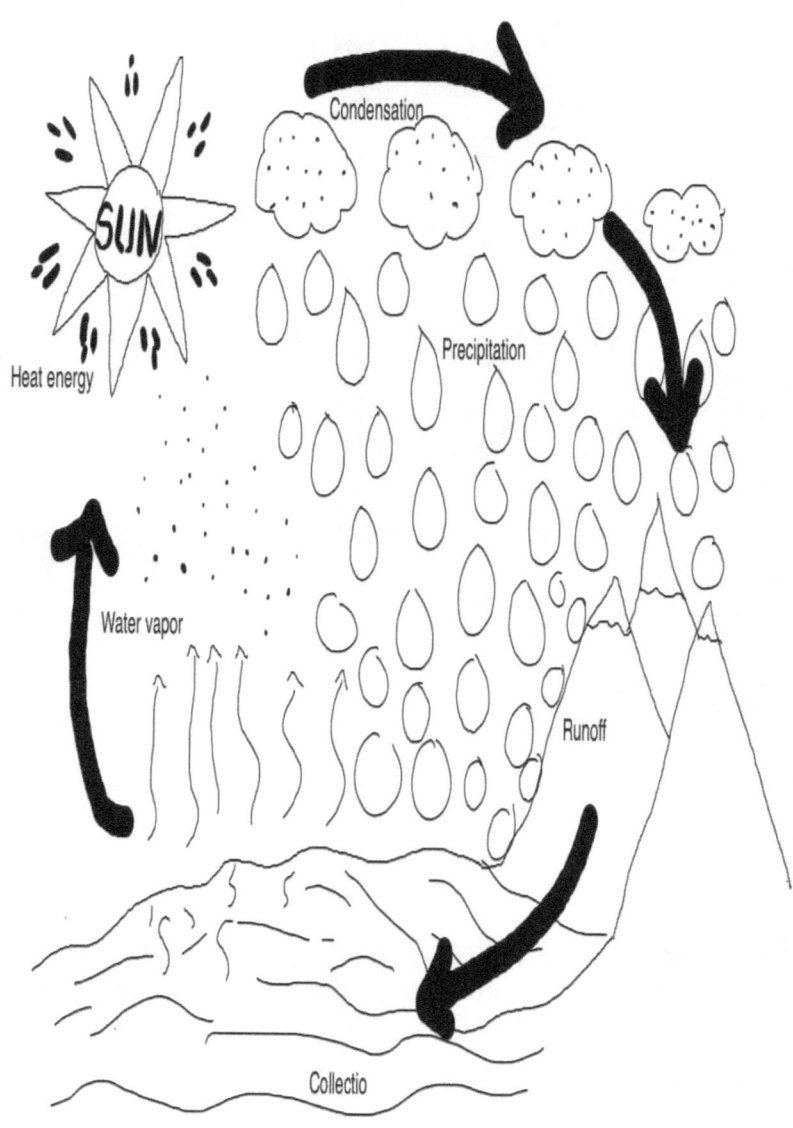

Plants must take in water from their roots in order to grow. Plants need water to make their own food and grow. Too much water or too little water could cause the plant to die.

Animals like you and me use water to wash and keep cool.

Tamika K. Fordham

We drink water to stay alive.

The water on Earth today is the same water that's been here for nearly 5 billion years.

Earth is the only planet known with living things and water.

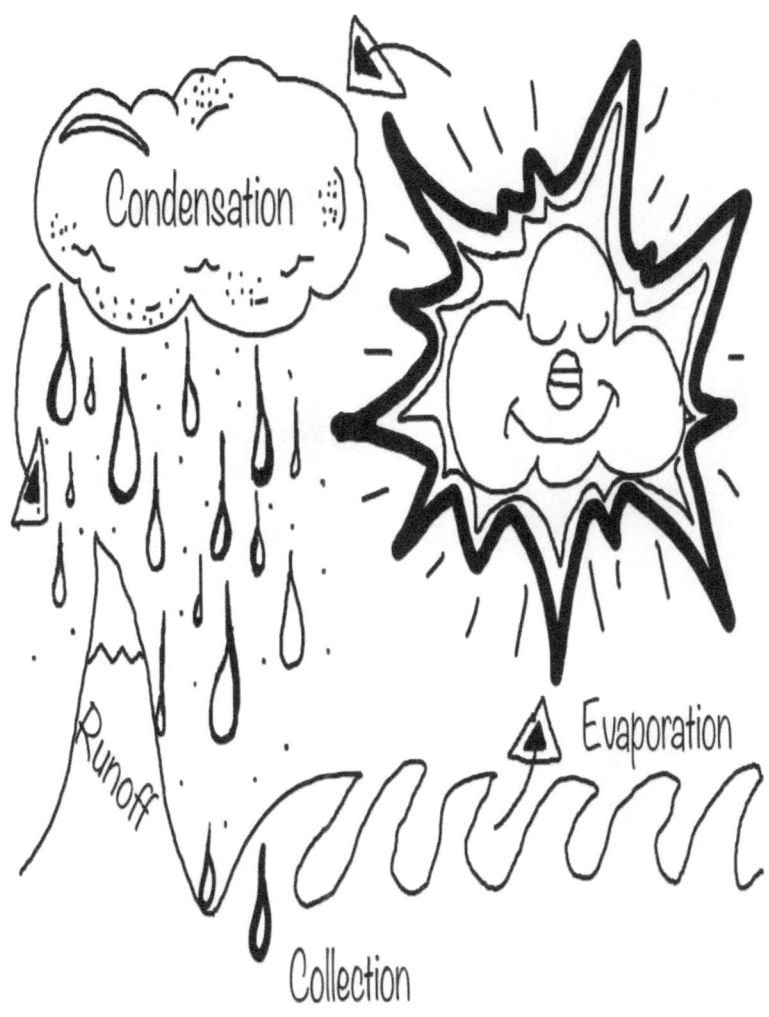

When precipitation falls on land surfaces, it attempts to run back to the ocean or lakes as runoff. If it falls from the clouds back to the ocean, it is called collection.

Tamika K. Fordham

Teach Me the Water Cycle and Weather

## ABOUT THE AUTHOR

My name is Tamika K. Fordham, wife of Kendrick Fordham and mother of Kennedy Fordham. I was selected as the 2017 NAACP Teacher of the Year, 2016-2017 District Teacher of the Year for Calhoun County Public Schools and as 2017 STAR Teacher.

I graduated with honors from South Carolina State University and earned my Masters of Arts in Teaching in Early Childhood. I also graduated with honors from Liberty University and received my Educational Specialist Degree in Leadership. I served in the United States Army Reserve throughout my college years and most of my teaching career. I became a highly qualified early childhood generalist teacher when I earned my National Board Certification. I currently teach science and have taught all subjects!

I give all of my praise to Him from above!!

www.ingramcontent.com/pod-product-compliance
Lightning Source LLC
Chambersburg PA
CBHW031601210526
45464CB00003B/1382